meet the sabertooth tiger!

The Fossil Files

pamela pettyfeather

For BJ and other curious and clever kids

contents

introduction: what was a sabertooth tiger?

Did you know there was once a cat with teeth longer than your whole hand?

Imagine spotting a giant predator lurking in the tall Ice Age grass. It's about the size of a modern lion—but thicker, heavier, and way more dangerous. As it steps closer, you spot its most famous feature: two huge, curved fangs, sharper than knives and longer than bananas.

Meet the **saber-toothed cat**—one of the most famous predators that ever walked the Earth!

Saber-toothed cats weren't just bigger versions of today's lions or tigers. They were a whole different type of cat, built for a world full of mammoths, giant sloths, and other enormous creatures. With their powerful bodies and deadly teeth, they ruled the ancient Americas for millions of years.

But then... they disappeared.

Why did one of the fiercest hunters in history vanish forever?

Scientists are still piecing together the clues, using fossils, bones, and even tar pits to uncover the secrets of this legendary beast. And you're about to join the mission!

Get ready to explore the incredible world of the saber-toothed cat—how it lived, how it hunted, how it vanished, and how it still teaches us important lessons today.

Are you ready to sink your teeth into this adventure?

Let's go!

Sabertooth Secrets

- **Smilodon's front legs were super strong–to wrestle giant prey to the ground.**
- **Their jaws could open nearly 120 degrees wide–almost double a modern lion's!**
- **Their fangs were surprisingly delicate–perfect for slicing, but not smashing.**

not just a giant kitty

. . .

IF YOU'VE EVER HEARD of a **sabertooth or saber-toothed tiger**, you might picture a giant version of a modern tiger with cartoonishly long teeth.

But here's the truth: The sabertooth tiger wasn't a tiger. It wasn't a lion. And it definitely wasn't a giant house cat. It was something completely unique—its own incredible, extinct species. Let's meet the real animal behind the myth.

meet *smilodon*

Scientists call the saber-toothed tiger **Smilodon** (SMY-luh-don). The name means "knife tooth," and once you've seen its fangs, you'll know why. There were several types of saber-toothed cats, but **Smilodon fatalis** was one of the most famous—and the one most fossils come from. Smilodon lived in the Americas during the last Ice Age, and it was one of the top predators of its time.

It was **shorter and stockier** than today's big cats. Its chest and shoulders were heavily muscled—perfect for grabbing and holding onto giant prey.

If a lion is a sprinter, Smilodon was a **grappler.**

what made smilodon so different?

Let's start with the obvious: **Those teeth.** Smilodon had two huge canine teeth—longer than your hand. They were **thin, curved, and sharp**—like steak knives. They weren't made for chewing or crunching bones—they were made to **slice**

But there's more: Smilodon could open its jaws nearly **120 degrees wide**—much wider than any living cat. Its **front legs** were thicker and stronger than any modern feline. It had a **short tail** (unlike lions or cheetahs), which made it better at leaping than balancing.

In short: Smilodon was **not** built for chasing prey across open fields. It was built for **ambush.**

Cat	Time Period	Size	Teeth	Hunting Style	Cool Fact
Saber-Toothed Cat (Smilodon)	Extinct (~10,000 years ago)	4 ft tall, 600+ lbs	Two huge saber teeth (up to 7 in!)	Ambush and wrestle	Could open its jaws nearly 120° wide!
Tiger	Modern	Up to 3.5 ft tall, 600+ lbs	Long, sharp canine teeth	Solo stalk and pounce	Biggest wild cat alive today
Lion	Modern	Up to 4 ft tall, 500 lbs	Strong canine teeth	Group (pride) hunter	Only big cat that lives in large groups
Cheetah	Modern	3 ft tall, 100–150 lbs	Smaller, sharp teeth	Chase at high speed	Fastest land animal (70 mph!)
House Cat	Modern	~10 inches tall, 10–15 lbs	Tiny sharp teeth	Pounce on toys (and bugs!)	Shares a distant ancestor with Smilodon!

it's not a tiger, and not a lion

Let's clear up some common confusion: The name **"saber-toothed tiger"** is catchy—but **Smilodon wasn't a tiger at all.** It's only distantly related to modern big cats like tigers and lions. Scientists call it a **"machairodont"**—a separate branch of the cat family that's now extinct.

So why do we still use the nickname? Because it helps people picture what Smilodon *kind of* looked like—even if the name isn't quite right. Think of it like a nickname that stuck. Cool… but a little misleading.

family tree check-in

Here's where Smilodon fits:

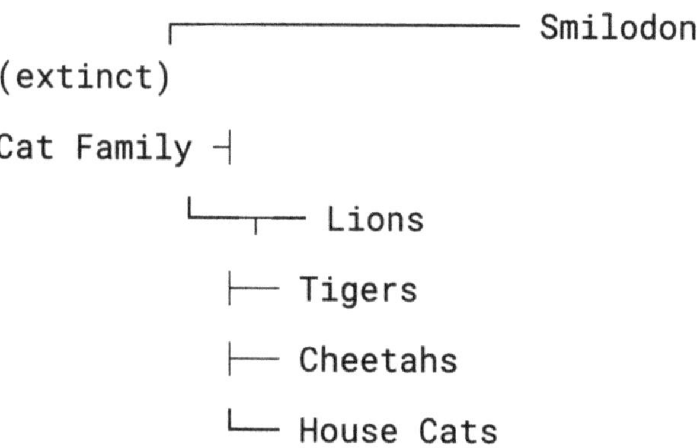

Modern cats and Smilodon share a distant common ancestor—but they evolved in different directions.

Did You Know?

- Smilodon's saber teeth were fragile –they could break if misused
- Some Smilodon fossils show healed injuries, suggesting they may have lived in groups
- It's one of the most common fossils found in the La Brea Tar Pits
- Their front limbs were so strong they may have pinned prey before delivering a bite

sabertooth color & compare

Label the key differences. Answers in the back of the book.

Next up: Where did Smilodon live? Who did it share the Ice Age with? And what did its world really look like?

Let's step back in time...

life in the ice age

. . .

LONG BEFORE CITIES, highways, or even farms, the Earth looked like a different planet. This was the **Pleistocene Epoch**—but you can call it the **Ice Age**.

Thick sheets of ice covered much of the planet. Temperatures were colder, winters were longer, and only the toughest creatures survived. But make no mistake: it wasn't just endless snow and silence. The Ice Age world was **teeming with life.**

what was the ice age, exactly?

The Ice Age lasted from about **2.5 million to 10,000 years ago**. During that time, gigantic **glaciers** covered huge parts of North America, Europe, and Asia. The **Earth's climate went through cycles**—sometimes colder, sometimes slightly warmer. **Animals evolved** to survive the cold: thicker fur, longer tusks, stronger bodies. And **humans** were just starting to spread across the globe!

While saber-toothed cats didn't live on the glaciers themselves, they roamed the **edges of the frozen world**, where grasslands and forests provided the perfect setting for ambush hunting.

what did their world look like?

Forget jungles and deserts. Most saber-toothed cats lived in **open plains, brushy forests, and dry, chilly grasslands**. These places were filled with herds of giant animals—perfect prey for a patient predator. There were no highways. No fences. Just herds of animals stretching to the horizon.

It was a time of **giants**.

who shared the ice age world?

The saber-toothed cat wasn't alone. It shared its turf with some of the most amazing creatures to ever walk the Earth:

- **Woolly Mammoth:** Giant elephant with long, curved tusks and shaggy hair

- **Giant Ground Sloth:** Slow, plant-eating tank of muscle, sometimes over 10 feet tall
- **Ice Age Horses:** Smaller than today's horses, but quick and plentiful—Smilodon snacks!
- **Giant Bison:** Massive buffalo with long horns
- **Camels:** Yes, North America had Ice Age camels
- **Dire Wolves:** Competitor predators, pack hunters
- **Short-Faced Bear:** One of the largest carnivorous land mammals ever—bigger than a grizzly!

This world was dangerous, dynamic, and always changing. Every animal played a part in the food web. Every hunter had to fight to survive.

where did saber-toothed cats live?

Mostly in the **Americas**—North and South. From chilly forests in what's now California to grassy plains in Argentina, different species of Smilodon adapted to different environments. Today, many of their fossils have been found in places like:

- **California** (La Brea Tar Pits)
- **Florida**
- **Texas**
- **South America**

Ice Age Neighbor Fun Facts

- A woolly mammoth's tusks could grow over 15 feet long
- The short-faced bear stood over 10 feet tall when on two legs
- The giant ground sloth had claws the size of bananas
- Some Ice Age animals were nearly wiped out by just a few degrees of warming

why the environment mattered

Saber-toothed cats depended on **open land** and **large prey animals** to thrive. When Ice Age climates changed, and forests replaced grasslands, life got harder for ambush hunters like Smilodon. When the prey vanished, so did the predators. In the Ice Age, environment = survival.

 Ice Age Animal Match-Up Game

Can you match the animal to its superpower?

Animal	Superpower
Saber-Toothed Tiger	A. Teeth as long as steak knives
Woolly Mammoth	B. Furry armor against the cold
Giant Ground Sloth	C. Could crush trees with its weight
Dire Wolf	D. Pack hunter with bone-crushing jaws
Ice Age Horse	E. Fast runner to escape predators

You can find the answers at the end of the book!

In the next chapter, we'll get even closer to Smilodon's most powerful abilities: Its body. Its bite. And its one-of-a-kind hunting style...

Get ready to meet the ultimate Ice Age predator.

the ultimate hunter

. . .

THE SABER-TOOTHED tiger wasn't the biggest animal in the Ice Age.

It wasn't the fastest. It didn't have horns, armor, or claws the size of swords. But it had something even more powerful: **A perfect plan.**

built to hunt

Smilodon wasn't built like a cheetah or even a modern lion. It was **shorter, heavier, and massively muscled**, especially in the front legs and shoulders. Think of it as a **four-legged wrestler**—low to the ground, solid, and powerful.

- **Front legs:** strong enough to tackle a moving bison
- **Neck and jaw:** built to support huge, slicing teeth
- **Back legs:** not for speed—but for explosive jumps
- **Tail:** short and stubby—not long and balancing like today's cats

Smilodon didn't chase. It waited. Then it **pounced.**

ambush, not chase

Modern big cats like cheetahs and lions rely on **speed** or **teamwork**. But Smilodon had a different strategy: It **hid in low grass or brush**, staying quiet and still. It waited for prey to come close. Then it **exploded into action**, tackling with powerful paws! It used its **saber teeth for a quick, deadly bite**—not to crush bones, but to cut deeply into soft areas like the neck or belly.

One wrong move, though—and those delicate teeth could break. That's why Smilodon was a hunter of **precision**, not power alone.

what was on the menu?

Smilodon mostly hunted **large Ice Age herbivores**, such as:

- Giant bison
- Ice Age horses
- Young mammoths and mastodons
- Giant ground sloths
- Camels

These animals were big, slow, and hard to bring down. But Smilodon had the strength—and strategy—to do it.

how strong was its bite?

Animal	Bite Force
Saber-Toothed Tiger	~1,000 PSI (strong, but not
African Lion	~650 PSI
Dire Wolf	~1,400 PSI
Short-Faced Bear	~1,600 PSI
Nile Crocodile	Over 3,500 PSI (!)

(PSI = pounds per square inch)

Even though Smilodon didn't have the strongest bite, its **entire body was designed to put it in the perfect position** to strike fast and finish the hunt.

Ice Age Predator Challenge!

Imagine you are a saber-toothed tiger. You see a young mammoth ahead. What do you do?

Circle your choices:

- **Wait and stay hidden OR Charge from far away**
- **Use your strength to tackle OR Try to sneak attack from behind**
- **Go for the neck with saber teeth OR Use your claws only**

Next up: Even the fiercest hunters couldn't survive forever. Find out what led to Smilodon's extinction—and what lessons it left behind.

why the saber-toothed cat disappeared

. . .

FOR MILLIONS OF YEARS, the saber-toothed tiger ruled its world.

With massive muscles, curved fangs, and a perfect ambush strategy, Smilodon was one of the most powerful predators of all time. Then—**in a blink of evolutionary time**—it was gone.

So what happened? Let's piece together the clues.

the world got warmer

Around **10,000 years ago**, the Ice Age came to an end. **Glaciers melted. Sea levels rose. Forests replaced open plains.** For animals that depended on wide, grassy hunting grounds, this was bad news. The saber-toothed tiger's environment—the one it was perfectly built for—**began to vanish.**

where did the prey go?

With warmer weather came big changes in vegetation. Grasses and cold-loving plants disappeared. Forests spread. And with them, **huge herds of prey animals began to shrink**.

- Bison populations dropped
- Horses migrated or died out
- Giant sloths and camels disappeared in many regions

Smilodon didn't just lose its hunting grounds—it lost **its food.**

then came humans

At the same time, **early humans** were spreading across North and South America. They hunted mammoths, horses, and bison —the same animals Smilodon relied on. They may have **competed** with saber-toothed cats for territory, and even may have hunted Smilodon itself!

Humans had something Smilodon didn't: **Teamwork. Tools. Fire.**

No saber-toothed cat—no matter how strong—could outmatch a group of humans with spears and human-tier strategy.

a deadly combo

Saber-toothed cats didn't disappear from **just one cause**. They vanished because **everything changed all at once**:

- The climate warmed
- Their habitat shifted
- Their prey disappeared
- A new predator (humans) arrived

Smilodon was built for a world that no longer existed.

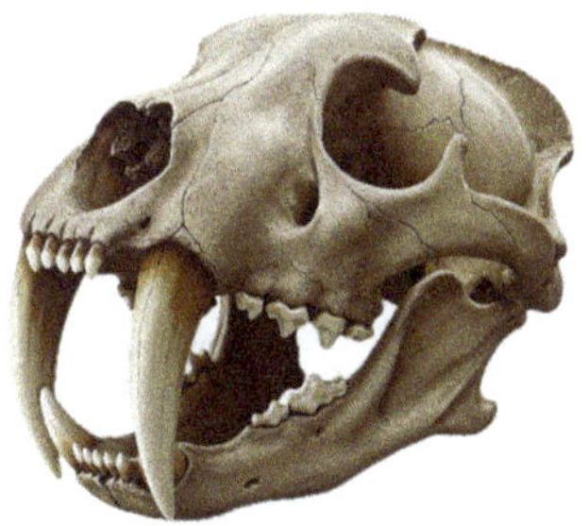

Did You Know?

- Over 70% of North America's large mammals went extinct at the end of the Ice Age
- Smilodon was one of the last saber-toothed cats to survive
- Some fossils show evidence of starvation and stress–clear signs that times were tough

what can we learn from smilodon's extinction?

The saber-toothed tiger reminds us that:

- Even the strongest animals can vanish
- The world can change faster than species can adapt
- Everything in an ecosystem is connected—plants, prey, predators, and people

Be an Extinction Detective!

What do YOU think was the biggest reason Smilodon disappeared? Circle one–or more:

- **Climate change**
- **Human hunting**
- **Loss of prey animals**
- **Habitat loss**
- **Bad luck?**

Now explain your theory on the next page: "I think Smilodon went extinct because..."

Next up: How do we know all of this? Let's dig into fossils, tar pits, and the clues Smilodon left behind...

fossils, tar pits, and amazing finds

. . .

THE SABER-TOOTHED tiger may be long gone… But it left behind something incredible:

Bones. Teeth. Clues.

And thanks to those clues, we know more about Smilodon than almost any other Ice Age predator.

How?

Let's take a closer look—into the ground, the tar, and the science.

what is a fossil?

A **fossil** is what's left behind when part of a living thing—like a bone, tooth, or shell—gets buried and slowly turns to stone. Over **thousands or even millions of years**, minerals replace the original material. It's like nature making a stone copy!

There are also other kinds of fossils:

- **Footprints**
- **Burrows**
- **Poop** (yes, really—it's called **coprolite**!)
- Even **fur and DNA** under the right conditions

For Smilodon, **bones and teeth** are the most common fossils—but what a story they tell!

what's so special about tar pits?

In Los Angeles, California, there's a sticky, stinky place called the **La Brea Tar Pits**. Thousands of years ago, it was a natural trap.

- **Bubbling asphalt** oozed from the ground
- Large animals—like bison and mammoths—got stuck

- Predators (like Smilodon) rushed in to attack…
- And got stuck too!

Tar preserved bones **better than almost anywhere else on Earth.** It's like a time capsule from the Ice Age.

what fossils tell us about smilodon

At La Brea alone, scientists have found:

- Over **2,000 individual saber-toothed cats**
- Bones showing **injuries** and **healing** (proof they may have lived in groups!)
- Teeth with wear patterns that show what they ate
- Skulls and jaws that prove their **bite** was powerful—but not built for crushing bone

From these fossils, we know:

- How big Smilodon was
- What it likely hunted
- That it was **more social** than scientists once thought
- And that it struggled during the end of the Ice Age

Did You Know?

- **La Brea Tar Pits have preserved over 3.5 million fossils**
- **Smilodon fossils are so common there, it's nicknamed the "Tar Cat"**
- **Scientists still dig up new bones there every year**
- **A single Smilodon skull can teach scientists over 20 different things about the animal's life**

how scientists study fossils

When paleontologists find a fossil, they:

1. Carefully dig and brush it out of the ground
2. Record its position and location
3. Take it to a lab for cleaning, measuring, and scanning
4. Compare it to other fossils and modern animals

5. Build **3D models, reconstructions**, and **theories**!

Some use **CT scanners**, **carbon dating**, or even **ancient DNA** to uncover more secrets. Smilodon may be extinct, but with today's tools, it feels more alive than ever.

Be a Fossil Detective!

You just found a saber-toothed cat jaw in the dirt. What questions would you ask? What clues would you look for? Pick 3:

- **What kind of teeth are these? (Sharp? Worn down?)**
- **How big is it?**
- **Is the jaw broken—or healed?**
- **What other fossils were nearby?**
- **Are there bugs in the bones?**

Next up: What can ancient bugs trapped in tar tell us about saber-toothed cats? More than you'd think—get ready for some tiny but mighty clues!

bugs in the bones!

. . .

SABER-TOOTHED CATS MAY BE EXTINCT, but they left behind a trail of clues—and **some of those clues are bug-sized!**

That's right. Some of the best secrets about Smilodon's world were discovered thanks to **ancient insects**.

How?

Let's dig in.

fossilized bugs = tiny time machines

When scientists dig up fossil bones, they sometimes find more than teeth and claws. Buried in the same layers of Earth, they discover…

- **Beetles**
- **Midges**
- **Lice**
- Even **fossilized bug poop** (yes, that's a thing!)

These ancient insects help tell the story of **what the Ice Age was like**—and how saber-toothed tigers survived (or didn't).

what bugs can tell us

- **Cold-loving beetles** = The climate was still freezing
- **Plant-eating bugs** = There were meadows, not forests
- **Carrion beetles** = Something big died nearby (maybe a Smilodon meal?)
- **Fur lice or ticks** = These may have lived on Smilodon itself!

Think of insects as **tiny reporters**. They don't roar or growl…

but they leave behind clues that help scientists understand the bigger picture.

Did You Know?

- **Scientists have found over 100 species of Ice Age insects at La Brea!**
- **Some Ice Age beetles were the ancestors of beetles still living today**
- **Insects are used to "date" fossils, just like detective tools at a crime scene**

science spotlight: la brea's bug box

At the La Brea Tar Pits in California, where **hundreds of saber-toothed cat fossils** have been found, scientists also discovered:

- **Tiny beetle shells** preserved in tar

- **Pollen stuck to insects**
- Even **fossilized flies** that once swarmed around the bodies of dead animals

These buggy bits help paleontologists figure out what the weather was like, how long an animal had been dead, and what kind of plants grew nearby—all from **just a few millimeters of bug parts**!

Be a Bug Detective!

Look around your home, yard, or classroom and find a (living) bug. Observe it carefully, and ask:

- **What might it tell a future scientist about the environment?**
- **What does it eat?**
- **Where does it live?**
- **How could you preserve its story?**

Write a Bug Report about your find here.

You're one step closer to being a fossil detective!

Next up: What can a saber-toothed tiger teach us today? It's more than just a cool predator—it's a symbol of survival, extinction… and the future.

what we've learned

. . .

THE SABER-TOOTHED TIGER IS EXTINCT. Its teeth don't glint in the sun. Its paws don't thunder across the grass. Its roar (if it had one) hasn't been heard for 10,000 years.

But its story? **That's still alive.** And it's teaching us more than ever.

why smilodon still matters

Scientists don't just study fossils for fun (though it *is* pretty fun). They study animals like Smilodon because:

- They help us understand **how life responds to change**
- They reveal how **ecosystems rise and fall**
- They remind us that even the strongest creatures can disappear

Saber-toothed cats were **apex predators**—at the top of their food chain. But they vanished when the Ice Age ended, their prey disappeared, and humans spread across the land. No predator—no matter how powerful—can survive when its world collapses.

what the past tells us about today

Here's the big picture:

- When **climates shift**, some animals can't keep up
- When **habitats vanish**, predators lose places to hunt and hide
- When **species go extinct**, it affects *everything*—even us

Sound familiar?

Today, the Earth is warming again. Forests are shrinking. Big cats like **tigers, lions, leopards, and cheetahs** are in danger.

And once a species is gone—just like Smilodon—it's gone for good.

the cats of today

You won't find saber-toothed tigers in the wild. But their modern cousins are. And they all need **space, food, and protection.** Just like Smilodon did.

- **Tigers:** fewer than 4,000 left in the wild
- **Lions:** pushed out of 90% of their historical territory
- **Snow leopards:** ghost cats of the mountains, extremely rare

Did You Know?

- **Every big cat alive today is classified as vulnerable, endangered, or critically endangered**
- **Some cheetah populations have fewer than 50 adults**
- **In the last 100 years, we've lost three species of wild cats forever**
- **Conservation scientists use fossil records to help plan for the future**

adopt a modern big cat

Pick one living big cat and "adopt" it—not with money, but with curiosity and care!

1. Learn where it lives
2. Discover what it eats
3. Find out what threatens it
4. Explore one way people are helping protect it
5. Make a poster or mini-presentation to share with your class, family, or friends!

the past helps protect the future

The saber-toothed tiger may be gone—but its **story is a warning and a guide.**

- Change is fast.
- Extinction is real.
- Nature is connected—from the tiniest bug to the top predator.

And the best part?

You're now part of that story.

You're a fossil detective.

A science explorer.

A voice for the animals of yesterday—and today.

Let's make it official.

Make a Wildlife Hero Pledge

"I promise to learn about animals, protect nature when I can, and speak up for creatures that can't speak for themselves."

Sign your name: ____________________

Date: ________________

bonus coloring pages

puzzle answers

chapter 1: sabertooth color & compare

tail length, front muscles, jaw size, teeth

chapter 2: ice age match-up

1–A, 2–B, 3–C, 4–D, 5–E

more fossil files

Unearth the Giants of the Past!

Step back in time and meet the amazing giants that once ruled the Earth! ***The Fossil Files*** is a kid-friendly educational series that digs up the most amazing prehistoric predators and Ice Age icons and brings them roaring back to life—one fossil at a time.

you decide the next fossil file!

Do you have a favorite extinct creature? Scan the QR code below to tell us what animal we should dig into next!

While you're there, join the **Fossil Files mailing list** for awesome extras—like puzzles, games, fun facts, and early sneak peeks at upcoming books!

who's your ice age twin?

Find out which Ice Age creature you're most like!

Take this fun quiz and unlock your free *Fossil Files Activity Pack!*
Scan the QR code above (or tap the image below)!

books for curious & clever kids

global explorations. bold questions. weird holidays.

Welcome to the *Books for Curious & Clever Kids series*—a collection of lively nonfiction books that bring world traditions, history, and culture to life for kids ages 8–12!

These books are designed for the endlessly curious and the slightly-too-clever kids who always ask: "Wait… Why do we do that?" "Where did that come from?" "Do other people do that, too?!"

Each book in the series explores a major holiday's surprising history and global story with Kid-friendly activities.

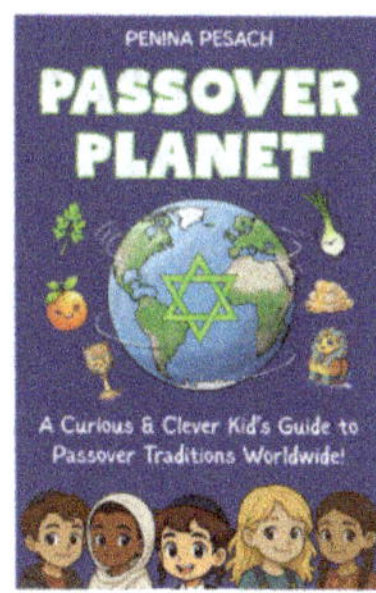

easter unscrambled for curious & clever kids

Why do we hunt for eggs on Easter? Where did the Easter Bunny come from? And what do ancient gods and chocolate bunnies have in common? ***Easter Unscrambled*** is the perfect blend of fascinating facts and hands-on fun — ideal for kids ages 7-12 who love puzzles, history, and surprises. This interactive book takes readers on a 4,000-year adventure through the origins of Easter — from ancient gods and springtime festivals to Easter egg hunts and chocolate bunnies.

---

halloween unscrambled for curious & clever kids

Why do we carve pumpkins? Where did trick-or-treating really come from? And what does a 2,000-year-old Celtic festival have to do with your Halloween candy? ***Halloween Unscrambled*** is a fascinating journey through history that reveals the surprising true story of how Halloween became the holiday we celebrate today.

From ancient Celtic bonfires to Mexican Día de los Muertos altars, discover how Halloween traditions traveled around the globe and evolved over centuries!

passover planet: a curious & clever kids' guide to passover traditions worldwide!

Frogs. Pharaohs. Pancakes. Passover Like You've Never Seen It Before! Get ready to travel through **3,000 years of Passover celebrations**—from **ancient Egypt to modern space stations**, from Moroccan mimouna feasts to Ecuadorian chocolate charoset! In *Passover Planet*, kids will **1)** Discover global Passover traditions from **Ethiopia, Persia, Poland, Iraq, South America** & more **2)** Learn about cool customs like scallion fights and Afikoman ransoms **3)** Explore **Seders with matzah pizza**, spicy charoset, and teff flatbread…

---

may day unscrambled for curious & clever kids

Why do people dance around poles in spring… while others wave protest signs? From leafy legends to labor marches, ***May Day Unscrambled*** unravels one of the world's most surprising holidays! Kids will travel through time and across the globe as they explore: 1) **Ancient spring festivals** like Beltane and Walpurgisnacht **2) British traditions like maypole dancing and the May Queen** 3) The rise of **International Workers' Day** and the fight for fair labor and 4) Global May Day celebrations in **France, Finland, Hawaii**, and more!

christmas unscrambled for curious & clever kids

Why do we hang stockings, bake gingerbread, and wait for a man in a red suit? ***Christmas Unscrambled*** peeks behind the traditions to reveal the history, legends, and global customs that shaped the holiday. Kids will explore: 1) Solstice festivals like Saturnalia and Yule 2)The origins of Santa Claus and gift-giving 3) Holiday foods, songs, and symbols around the world 4) Christmas celebrations from Europe, Africa, Asia, and the Americas

fraudulent folklore: a cryptid coloring, fun facts & creative writing adventure

What's the truth behind Bigfoot's big feet, Nessie's blurry photos, and Unicorn horns that turned out to be whale teeth? ***Fraudulent Folklore*** is the perfect mix of fascinating facts, funny legends, and creative fun — made for kids ages 8–12 who love monsters, myths, and using their imagination. With 25 silly cryptid "bios," single-sided coloring pages, and story-writing prompts for each creature, this interactive book invites kids to read, color, and create their own monster legends.

READ THE LEGENDS. COLOR THE CREATURES. CREATE YOUR OWN MONSTER STORIES—ALL IN ONE HILARIOUS AND EDUCATIONAL ADVENTURE!

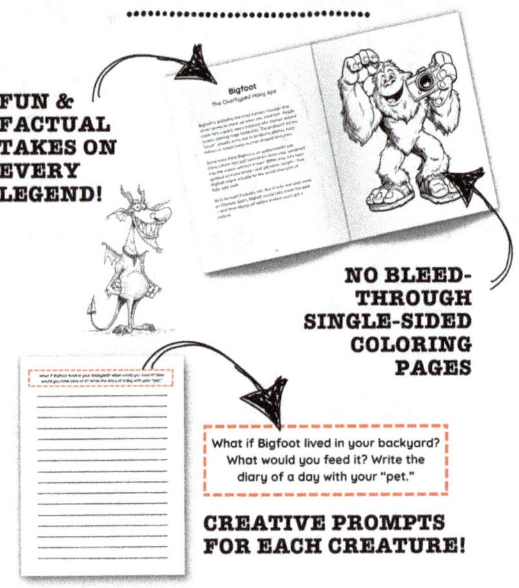

+ PRINTABLE PUZZLES

www.ingramcontent.com/pod-product-compliance
Lightning Source LLC
LaVergne TN
LVHW050542100826
845148LV00002B/653

9798899680175